# The Journey through Developmental Mathematics

## *An Analysis of Decision Making Regarding Attitude and Time Management*

Dr. Darrin Berkley

INFINITY
PUBLISHING

ISBN 978-0-7414-6910-6 Paperback
ISBN 978-0-7414-6911-3 eBook

Printed in the United States of America

Published February 2023

INFINITY PUBLISHING

Toll-free (877) BUY BOOK
Local Phone (610) 941-9999
Fax (610) 941-9959
Info@buybooksontheweb.com
www.buybooksontheweb.com

The goal of this pamphlet is to give developmental mathematics students some concrete advice regarding their "journey" through the world of mathematics. This journey is made up of the many individual decisions students make (or do not make) as they strive for success in developmental mathematics. Generally, students begin in different places on the journey, starting with the first course of developmental mathematics (Basic Mathematics) and continuing on to the last course (Intermediate Algebra). Throughout the journey there will be many occurrences of both joy and frustration. A challenge for the students is channeling their frustration.

In the math courses that I teach, students are asked about their past and present mathematical experiences. This allows me to carefully analyze the decisions that have had the greatest impact on their journey in developmental mathematics. Community college students deal with many issues impacting their academic abilities; consequently, this pamphlet will not address every issue that students face. However, this pamphlet will discuss the most common issues faced by developmental mathematics students.

The pamphlet should be read sequentially. Students should address issues in the first chapter before reading the second chapter. This may require additional communication with instructors, tutors, employers, family or school administrators.

This pamphlet will read as if you are a student sitting in my office. Since you are reading this, it is likely that you would like to do better in your mathematics courses. I'm not going to provide you with research studies in mathematics education or expect any complex tasks from you. I am going to give simple advice as well as determine the likelihood that the advice will be followed. As with any problem, it will probably require a significant amount of your time to solve. As you are going through the pamphlet, analyze the decisions that you make as well as the consequences resulting from your actions. Finally, if you are serious about succeeding at this, believe in yourself!

Let's get to work!

## *Chapter 1: Adjustment in Attitude*

Let's try to understand your mindset. Most students that are enrolled in developmental mathematics either do not like mathematics or take no interest in mathematics. What comes to mind when you think about the course in which you are enrolled? If you don't like mathematics, try to understand why you feel that way.

One reason might be that you haven't been successful in mathematics in the past. Why not? First, do you truly believe that if you give your very best, there is a possibility that you could be successful? Don't tell yourself that you can't do it unless you've given everything you got. Also, think about those times when you *have* been successful. What did you do to succeed then?

If you believe that you can do the work, are you serious about maximizing your success? What do I mean? I am referring to the urgency of completing this journey. How important is it to you? Will your behavior illustrate the importance of pushing yourself through despite potential times of frustration? How will you deal with this? Whatever you do, don't quit!

Your success throughout this journey will likely depend on the decisions you make when you are frustrated. If you're quick-tempered, do you have a plan for controlling your temper? Speaking for all who have completed this journey, I can tell you there *will* be times of frustration. Understand that you can overcome these times. Let's try to better understand your frustration.

We need to go deeper than "I've never been good at math." We are not going to dwell on the past. Let's try to go forward with the skills that you have. As an instructor, the only concern I would have at this point is whether or not you meet the required prerequisites of the class. In other words, do you have the foundational skills necessary for success? If not, you may want to change to a lower-level developmental math course. If you have the skills but have forgotten what you have learned because you haven't taken math in several years, ask the instructor or the mathematics department if you can take an old final exam from the course prior to the one in which you are enrolled. Even though taking a prerequisite course may cost you more money, it will be beneficial in the long run for your basic foundational knowledge. Remember, you should be placed in the course that best represents the mathematical skills that you have. You will also gain more confidence by seeing that you may not be as bad at math as you think you are. Being placed in the appropriate math course will allow you to more accurately represent your ability and will likely decrease your frustration during your journey.

The next possible source of frustration may be that the explanations given by your instructor are unclear. Now you understand why having the prerequisite skills must be discussed first. If you don't have the necessary foundational knowledge, then it's possible that the explanations given by your instructor will always be unclear regardless of how many times he or she tries to explain the course content. Not understanding your instructor's explanations can be especially frustrating when you need clear explanations to complete homework or other assessments. You can solidify some of the explanations given by your instructor by asking questions or verifying that you understood the explanations.

Are you afraid to ask questions in class? Do you feel that asking questions will slow the class down? Put away your ego! Remember that you paid for the class. It's expected that you'll have questions. This is why you are enrolled in *developmental* mathematics. You need to *develop* your skills so that you're adequately prepared for college-level mathematics. You must ask questions so that you can better develop your mathematical skills. Most developmental mathematics curricula are designed so that students have time to ask questions. Anyway, your instructor manages the pacing of the course regardless of whether students ask a few questions or many questions. If too many questions are asked, the instructor may choose to answer questions during office hours or during the next class session. The instructor may also use email correspondence to answer questions. The instructor is also aware that a certain amount of content must be covered during the semester. So the challenge for the instructor is to find that balance between covering the course material and making sure that most students get their questions answered during the few hours of class time each week.

As an alternative to asking your instructor, you can use the Internet to get answers to your questions. If you don't understand a topic—for example, factoring—use a search engine such as Google or Yahoo and type in the word or phrase you're looking for. You'll likely get many links that you can use to help you answer your question. In fact, sometimes you can type in your exact question to get a response. Another option is to consider specific sites, such as Dr. Math, that might answer your questions. You may consider used books (ask your instructors—they generally have many of them) or CliffsNotes to answer your questions as well. You may try these options, but meeting with your instructors still proves beneficial.

Use your judgment. If you feel that you have too many questions to be addressed in class and you've exhausted all other possibilities, then you'll need to sacrifice some time and meet with your instructor during office hours. If the office hours given by your instructor don't fit in your schedule, then ask your instructor to meet with you at a different time. Although there are other people that you can turn to for help, it is important that you maintain good communication with your instructor. Before you meet with your instructor, make sure that you have completed all previous homework assignments in order to maximize your ability to understand your instructor's explanations. Also, make sure that you understand to the best of your ability what your instructor has already said in class. Be sure that your notes are 100% accurate. Allow the instructor to give the exact same explanation that was given to you in class. If this doesn't work, ask for an alternative explanation. Instructors often have multiple ways to explain the same idea.

What also may occur is that you might not get along with your instructor or you might simply believe that he or she doesn't have enough experience to clearly explain the course material that is causing you confusion. In that case, a new plan of action must be considered.

Is it possible to mend the relationship between you and your instructor? What exactly are the problems between you and your instructor? Remember that in order to resolve the conflict, you'll have to be clear when stating your problems to your instructor (or to a third party, if necessary). Before you speak with your instructor, understand any underlying issues that have contributed to the problems. Also, think about the decisions that you have made that may have contributed to the problems.

Understand that not all of the problems you may have are solvable. One such problem is a possible language barrier if the instructor's native language is not English. You have to decide during the first week of classes whether or not the problem is significant enough that it impacts learning.

A second problem that may not have a solution is the grading of your assessments. There is a difference between "hard" grading and "unfair" grading. If your instructor is a hard grader, he or she may have high expectations and want you to be fully prepared so that small mistakes don't carry over to other courses. Small mistakes can generally be fixed if you look over your work carefully. Instructors assign grades through the use of a grading rubric. However, some instructors don't make the grading rubrics for all assessments available to students. Feel free to ask your instructor how assessments are graded. Some rubrics do not award partial credit for wrong answers on certain assessments or individual questions. This holds true both for handwritten assessments and computer software programs.

How do you deal with an instructor who grades unfairly? First, understand that most instructors want you to succeed. (Otherwise, why teach?). Your instructor is not the enemy. An instructor typically would have no reason to "target" a student. Sometimes, the unfair grading is not always intentional if it indeed occurs. Instructors are not perfect: they will make honest mistakes in their grading from time to time. If you feel that you have been graded unfairly, do not let it go!

You must first trust that your instructor has the mathematical competence to grade your assessments fairly. If you find that you cannot have this trust, you may find yourself questioning every grade your instructor gives to you. This can create a toxic relationship between you and your instructor. In this case, it would be best if you reregister for the course with another instructor if possible.

In each of these scenarios, you must control your anger and keep your attitude in check. Strong, positive communication with your instructor is critical. Be sure that you reevaluate your assessment from both your point of view and from your instructor's point of view. Try to view your assessment through the eyes of your instructor. Was something unclear in your answer that might have caused your instructor to mark the answer incorrect? Did the instructor make a correction without actually deducting points? Understand the impact of the assessment on the overall course grade. Is it worth creating a strained relationship by saying something to your instructor simply out of anger? Don't let one incident impact the continuous communication that you should have with your instructor.

If your instructor concludes that your assessment was graded unfairly, then you should obviously receive the points back that you have lost. Do not take this personally. It is likely not the first time that an instructor has made such an error.

There are also instances when students don't realize that what they have done is incorrect. If this happens to you, ask your instructor to explain (either verbally or through the use of a grading rubric) why points were deducted on your assessment. If the explanation given for your grade remains unclear, you can ask your instructor for an alternative explanation. If this fails, your next option is to consider having another instructor—such as one who you know from a previous class—take a look at your assessment just so you can get a second opinion. If no such instructor exists, then ask the department chair to give you an explanation for the grade you received. You can try to make a case for why you think you were graded unfairly, but do not be upset if the situation is not resolved in your favor.

In summary, be sure that you clearly understand the grading policies of your instructor. However, your main focus must be on *learning* the course content. An instructor can often tell which students genuinely want to learn and understand what they are being taught and which ones are just trying to get enough points to get a passing grade. If you have no interest in learning, then it will be difficult to succeed in any course, not just a developmental mathematics course. The challenge in having "an attitude to learn" is finding that math topic that sparks your curiosity. This represents another source of frustration for students: working hard to learn something that will never be used outside of the classroom. If a student doesn't see the usefulness of a particular topic, then the topic will likely be forgotten, which will create problems in future mathematics courses. So it is important for instructors to emphasize the importance of gaining a strong foundation in mathematics to alleviate possible frustration in the future.

It is probably true that if your major is art, you probably will have no use for the quadratic formula outside of the mathematics classroom. A future nurse doesn't recognize the importance of factoring in everyday life. When would I need to combine like terms as a business major? These are questions and statements that have come from developmental math students. At first glance, it's hard to see why these topics would be relevant to real life. Let's think a bit about why you are being taught mathematics.

I agree that at times you will be taught mathematics topics that you will never use. Remember, I am being honest as a mathematics instructor. But I could also make the case that very few courses are made up of topics that *all* students can use outside of the classroom. For example, do I need to know what a verb is to be a productive citizen? Do I need to know the periodic table of the elements? Do I need to know anything about the American Civil War in 1861? How will any of this help me now?

In practically all disciplines, we will always be taught more than we actually use. But we should all agree that it is better to have too much knowledge than not enough knowledge. Let's not look at mathematics as a set of discrete, disconnected topics. Instead consider mathematics as a language that all of us must use at some point in our lives. In thinking of language as a method of communication, you can probably think of many words that you would never use in real life. Understand that the students who are going through the developmental mathematics journey with you—those who sit next to you in class—come from all walks of life. Some want to be teachers, others want to be mathematicians, and some don't know what they want to be. The mathematics curriculum has to accommodate all of these students. This is by no means an easy task. The mathematics curriculum is not intended for any particular group. It is intended to develop basic knowledge that you are expected to know. Let's think about the skills that you strengthen other than mathematical ability. Why do you even need to take mathematics at all?

Some of the many uses of mathematics that are relevant to all students are the understanding and study of quantity, change, reasoning, structure, measurement, patterns, and risk. If you look at these topics carefully, you'll discover that each of these skills are used in the lives of all productive citizens. Many of these skills are learned indirectly and are absolutely essential to understand. This is why mathematics is taught at every level of schooling and why it must be understood on the college level in order to receive a college degree.

What else do you learn in the developmental mathematics classroom? It is important that you believe that skills can be learned during the journey in developmental mathematics. Since mathematics is often viewed as a "black-and-white" discipline, there is little room for error. Again, this creates frustration for students. You probably have experienced times when you have spent hours working on one problem only to find out that it was wrong due to a very small error. You may also have spent long hours completing an assignment only to key the answers into the computer mathematics software incorrectly and receive no credit. This kind of incident is more common in mathematics and science than in other disciplines. Of course, students generally don't question the importance of accuracy unless it has an impact on their grade. Whether in school or in the workplace, when tasks are to be completed, one expects the task to be completed accurately. You may get frustrated at an instructor when he or she takes of points for inaccuracy, but consider an inaccuracy in your paycheck—to your disadvantage—and think about the importance of accuracy.

Another important skill you learn is how to take a rule and then apply it to unknown situations. This is a skill that you probably already have. Mathematics can be used to strengthen this skill. You can think of rules as those that have been given to you by parents, those that are decreed by law, those given by your employer, and many more. These are rules that you may not understand completely, but you try your best to apply them. You are given many rules in developmental mathematics, and you often have to apply multiple rules to a single problem. Not understanding one rule could have a significant impact on the entire problem. Simply trying to memorize all of the rules will not allow you to apply the rules in various contexts. This is an idea that is prevalent in problem solving. Word problems often serve as those "unknown" situations. Understanding how to apply rules can be a useful skill in real life as well. One of the most common expectations for most jobs is likely to be an ability to solve problems.

Mathematics can also improve your critical thinking ability. Critical thinking has many definitions, but some of the characteristics include determining relevant and irrelevant information, making accurate judgments based on the problem context, questioning assumptions, and using inductive and deductive reasoning. All of these characteristics are used outside of mathematics as well. Some argue that critical thinking is not used in developmental mathematics as often as it should, but there is no doubt that critical thinking plays a significant role in education in general as well as in mathematics.

What does mathematics offer that may be relevant to you? One skill that you strengthen is your patience. How much patience does it take to continue taking mathematics courses that do not count for credit? If you are enrolled in a developmental mathematics course, it is likely that your patience has been tested on numerous occasions. Some students find themselves losing confidence in their ability and losing patience as a result of poor results on tests or other assessments. The question arises as to what decisions students should make to address this issue. It is hard to convince students to remain patient when they continue to receive poor results over long periods of time. How much do you give before you just quit? This is a question that you may have asked when completing homework assignments. Some students quit if they can't solve a problem on the first attempt. Others make numerous tries before quitting. Ask yourself how you feel when you are in these situations. Do you try to get assistance when you struggle? Or do you start to tell yourself that you just can't do it?

How have you handled situations similar to this outside of the mathematics classroom? Did anyone or anything contribute during the times when you remained patient? Another good idea would be to read about other people who had to remain patient to achieve a goal. It can take a lot of hard work, but you can do it if you apply yourself and think about the end result. Remember that you are not by yourself. Whether they have an educational background or not, family and friends can play a supportive role in helping you maintain your patience.

Some students feel that if they can succeed in mathematics, then they can succeed in anything. Learning mathematics is a challenging task, but I'm sure you know that understanding mathematics is not impossible. Mathematics may not come to you as quickly as it comes to others, but it is not a race. Be patient and understand that with your failures can come great success. That success is having enough patience to complete the journey.

We have just discussed some of the reasons why mathematics should be important to you. Mathematics is all around you! Since most companies use formulas to some extent, understanding those formulas does not just affect those that devise the formulas—it will likely impact you as well since your life can be affected by the formulas. So do not have the attitude of "Somebody else needs to understand those formulas." Counting on someone else's knowledge when you could obtain that knowledge yourself can cost you thousands of dollars in the future. Many community college students change majors. You never know when you'll change your major to one in which mathematics will play a greater role.. Don't take your chances! Learn now when you have the opportunity. It may not present itself again.

So what if you choose not to learn mathematics? What are the consequences?

Well, one of the obvious consequences is that you will not get the degree that you worked hard to obtain. On many occasions, students put a great deal of time (not necessarily productive) in their mathematics courses. An associate degree (or any degree for that matter) requires a lot of dedication and drive. It will require you to do things that you do not want to do. Because of the importance of mathematics, it is a requirement for your degree. Colleges and universities expect students to be literate in mathematics and science to compete in a global economy.

Second, it is just a waste of valuable time. All of us believe that our time is valuable. If you're going to take a developmental mathematics course, why not give it your best the first time? There could be many answers to this question. In many cases, the student knows what needs to be done to succeed. The student must have the mindset to learn the mathematics content even if he or she does not have an interest in mathematics. Sometimes, students try to memorize the course material because it is easier to accomplish and saves time. The problem with that is that there is often too much material taught in a semester to memorize all of it. Also, just because you have memorized how to do a problem doesn't necessarily mean that you understand it. Memorization can get you through some courses, but it will eventually catch up to you if you choose not to learn. Memorization will save you time in the short term, but will likely result in a major loss of time in the long term.

Third, it is a waste of money. Even if you are on a needs-based scholarship or grant, someone's money is still wasted. It has an even bigger impact if you pay for your own education. Most students will admit that they don't have money that they can just waste or throw away. If you're getting help from parents, your parents may decide that you are not serious about your education and choose not to pay for your classes anymore, causing you to drop out of college.

Fourth, you may not get accepted into certain academic programs. Some programs require that you complete a certain number of mathematics courses with a minimum grade for acceptance. Remember that you are competing with many other students. Generally, the hardest-working and most knowledgeable students are likely to get into such programs.

Fifth, a failing grade can take a huge hit on your GPA. Often a certain GPA is required for acceptance into certain programs. Also, some employers will look at your GPA for employment purposes. At some point, you may consider applying to graduate school. Admissions departments consider GPA as a factor for acceptance into graduate programs.

Finally, what does a failing grade tell you about yourself? What does it tell others about you? Often it represents failure from different perspectives. For those who really don't understand what the "F" (or "D") means on your transcript, you still may have to answer to parents or other family members. Students who live with their parents may have to answer to them, which may not be a pleasant experience. Older students who fail may have to answer to their spouses because of the economic impact on the family. Sometimes, a degree may be a new requirement for a current job. Not having a degree may result in the loss of future opportunities, which could have serious life consequences. One way to avoid these consequences is to simply be a good student. In other words, just do what your mathematics instructor asks you to do.

As an instructor, I'm going to share some things you can do to prove that you are being a good student. If you are willing to be a good student, I expect you'll have success in developmental mathematics. Ask yourself if you are willing to commit to these tasks. If so, promise yourself that you will make the necessary sacrifices to complete the following tasks. Make no excuses for not completing the tasks. This will show that you are serious about completing your journey.

1. **Focus in Class**

   Stay awake; cell phones should be put away during class; come to class prepared to learn; keep side conversations to a minimum

2. **Come Prepared for Class**

   Textbook; several pencils (erasers); required calculator; notebook

3. **Adjust to the Pace of the Class**

   Student must make adjustments

4. **Allot Time to Do the Work**

   At least **2 hours** per day (14 hours per week) outside of class is necessary. This is essentially a part-time job.

5. **Handle Frustration Appropriately**

   Get the help you need. Speak with your instructor regarding **any** concerns. Use the math lab. Learn from your mistakes.

6. **Ask Questions**

   Put away your ego. If you knew it all, you wouldn't be here. Don't worry about what others think. If you don't know something, tomorrow may be too late to ask.

**7. Handle Transportation Issues**

Leave for class at least 10 minutes to allow time for traffic and parking. Plan on staying for the entire class.

**8. Complete Work in a Timely Fashion**

Respect the due dates. Late work is not accepted. "I forgot" is unacceptable.

**9. Understand Errors Made on All Assessments.**

Students are held to the highest standards regardless of ability level.

**10. Understand the Consequences of Your Actions**

Whether the results are good or bad, hold yourself accountable!

**11. Read All Assigned Sections in the Textbook**

These present the main ideas in the unit and cover the topics that will most likely be on your exam.

**12. Complete a Practice Exam**

If you have math anxiety, it is important that you simulate the actual exam and analyze any errors.

## *Chapter Summary:*

This chapter discussed the importance of having an attitude of learning. You must take the course seriously, and it should show in your actions and decisions throughout the journey. During this journey there will be times of frustration. How you handle your frustration will impact the journey. Some of the frustrations include previous negative results on mathematics assessments, unclear explanations by your instructor, language barriers between instructor and student, poor student-instructor relationship, and grading concerns. Another frustration that was addressed was the relevance of mathematics. In other words, why do you even need to take a mathematics course? Once the relevance of mathematics is recognized, then students must understand the consequences of not understanding mathematics. These consequences are not pleasant to face. You can avoid these negative consequences by being a good student and doing what you are asked to do. The next chapter assumes that your attitude of learning has been established, and we will move on to discuss the impact of time management.

## *Chapter 2: Time Management*

Now that you have made any necessary adjustment in your attitude toward developmental mathematics, we'll look at the amount of time you can devote to your coursework. The first step is to create a schedule for every day of the week including everything that occupies any of your time. Do not forget to include transportation time to work or school. As an estimate, look at your schedule to determine if you have a minimum of at least 14 hours per week for your mathematics course. Why 14 hours? Remember that you will have to consider the amount of time it takes for completing homework assignments (this includes some challenges that might occur in solving problems), reading a section of a textbook multiple times prior to class to maximize understanding, studying for examinations, completing of small projects, reviewing your notes, understanding computer software, attending tutoring sessions and office hours, reviewing previous graded assessments, and so on. Note that 14 hours per week is similar to a part-time job, but you may or may not need all those hours. It is better to overestimate the number of hours you need for your developmental mathematics course than to underestimate. This estimate will change depending on your mathematical ability. Also, if you have a tutor working with you from the first day of the course, this may minimize your time. Again, to be safe, I would be hesitant to enroll in a developmental course if you can't devote 14 hours per week to the class. If the mathematical topics are more challenging than you anticipate, if you have a poor student-instructor relationship, or if it has been a long time since you have taken a mathematics course, expect to devote at least 2 hours per day or more. Are you willing to put in this amount of time to pass your developmental mathematics course?

To get a better estimate of your time, you'll need to analyze both speed and accuracy when completing assignments. Time yourself when given an assignment. How long does it take for you to complete an assignment? What percentage of questions did you get correct? Generally, you are given the answers to your homework assignments so you can confirm that you have completed your homework assignments correctly. If you feel that it is taking you too long to complete assignments, then speak with your instructor to verify that you have a thorough understanding of the intermediate steps involved in solving problems. Have your instructor give you a problem and allow your instructor to watch you as you complete the problem. Remember that checking your work will decrease speed but increase accuracy. It is strongly recommended that you take the extra time to check answers when possible. This will allow you to minimize the number of careless mistakes that you might make.

Suppose you don't have this amount of time to devote to your mathematics course. What is your next course of action? Because time is so important, I would recommend that you reregister for the course when you have this amount of time to devote to the course.

One activity that often takes a significant amount of time is your work schedule. This is a tricky topic to address. As a community college student, most students must work. Often there is no other option because many students pay for their education out of their own pocket. It's hard to tell students to consider working fewer hours because often they need the money to support their family. The question often arises, what is more important, work or school? Sometimes, one or the other has to be given less time, which can result in ineffectiveness. Many students consider work more important because without work they can no longer pay for school. With the impact of the bad economy, it is hard to turn down opportunities to work more hours. So some students decide to take the risk of working long hours and devoting little time to their mathematics courses. If it is possible, working fewer hours can give students additional time to complete their mathematics assignments. Other factors that may affect the decision to work fewer hours are whether or not homework can be completed on the job, the flexibility of the work schedule, and the student's overall financial situation.

Students may also allot more time to their mathematics course by enrolling in fewer courses. Many students underestimate the amount of time required for particular courses. As with your mathematics course, speak with the instructors of the other courses in which you are enrolled before the courses begin. Also speak with some students who have already taken these courses. This will give you a better idea as to how much time you'll need for the courses. This will also help you revise your schedule so that you know if you have enough time to devote to all your courses.

A challenge that many students face is a rush to graduate. Students aren't always patient or in some cases they just really want to get into certain programs at a particular time. There is an urgency to complete courses as fast as possible, especially when students are enrolled in school part-time. Sometimes students get an increase in pay once they get a degree. How do you convince a student that he or she is enrolled in too many classes? Students do get recommendations from the advising department, but some students do not necessarily take the advice of the advisors. Ask yourself if there should be a rush to graduate. What is your ultimate goal when you graduate? What opportunities does graduation give you? Will those opportunities always be there or must you move quickly to graduate? Answers to these questions will vary according to the individual. There will be a certain amount of risk that each student must consider when making the decision to take a significant number of credits.

The amount of time you have for your mathematics courses may also be affected by transportation. If you don't have a reliable means of transportation, you will likely be at a huge disadvantage. Not all places have convenient public transportation, and using a taxi to get back and forth to classes may not be affordable. Even though you could rely on others, remember that it is *your* education. If you are late or absent for a mathematics class, it can be very hard to recover. It will be you who suffers because of someone else. If you don't have a reliable vehicle, what can you do to get a vehicle (even if it is a used vehicle)? Can you buy a used (working) vehicle from a dealership or auction? Can you save some money or borrow some money to purchase a vehicle? If this is not feasible, can you use a family member's vehicle to travel to and from classes? Do you know someone who travels to school and leaves school at the same time that you do?

How does this impact time in your mathematics course? You may need to see your instructor or tutor outside of class time, which means that you might have to travel back to school for the appointment. If you don't own a computer and you're enrolled in a mathematics course that requires completion of online homework, you may need to spend several hours in the library—and you cannot guarantee what time you'll be finished. You may need to take advantage of study groups to prepare for your mathematics tests. Again, this would be hard to accomplish without a reliable source of transportation. Study groups can be beneficial in reducing the amount of time it takes to complete assessments or prepare for an exam.

Your family also can impact the amount of time that you can devote to your mathematics course. This can include caring for sick relatives, taking care of children, sharing a single vehicle with relatives, doing chores or tasks required by family members, and so on. If your family currently requires a lot of time, it may not be the best time for you to enroll in community college. Only you know the amount of time that your family members require, so I'll only bring to your attention the importance of your family. You may need to have a discussion with your family members about the importance of your schooling if you indeed want to be successful in your courses. They might allow you more time for your mathematics course and your other courses. Do you have babysitting in place for your children if needed? Do your children or any of your relatives have special needs that require a great deal of your time?

Finally, there could be a variety of personal issues that this pamphlet cannot address. If you have special needs or have to make frequent visits to the doctor, this all must be considered before considering taking a mathematics course. Each student has to determine what is most important in his or her life. The significance of these personal issues on a student's education can vary greatly. Personal issues can also include any addictions that could impact a student's life. What plan do you have to address these issues? If you can't avoid these issues, think of how you could minimize them.

A lot of factors have been discussed here that can impact your time. What is most important to you? What takes up most of your time? You will have a hierarchy of tasks. Determine the order of importance of these tasks. Where does school fit in this hierarchy? More specifically, where does learning (not memorizing) the topics in your developmental mathematics course fit into this hierarchy? Since most students who read this do not have an interest in mathematics, it may go lower in the hierarchy than other things. If this is true, it is also possible that it won't get enough of your time.

After a review of what takes up your time, the next question to ask is if you are willing to devote the time to your coursework. Some students don't have any of the issues discussed above. They simply say, "No, I am not going to do the work because I don't want to do it." Do you believe that you're just lazy when it comes to completing work in mathematics? These students would rather give their time to other tasks that they enjoy. They have never liked math and never will. They may also find that they don't want to work hard at something in which they think they won't succeed. The decisions that students are asked to make in this pamphlet *will* require some sacrifice. Think about the hierarchy you considered above and ask yourself why you created it in such a way. As with work, school is not always going to be fun. Think of the many things that you must do that are not fun. Hopefully, you are not deceived by reality TV. Their reality is not going to be your reality. Time has to be given for hard work that is not always immediately rewarding. Try to look forward to the potential benefits of completing the journey.

## *Chapter Summary:*

Students need time to succeed. Students can do well in their courses if they have enough time to complete all of the assignments. A very detailed schedule is essential for students to truly determine if they have enough time to give to their mathematics course. Tasks that have the greatest impact on the amount of time students have for their mathematics courses are number of hours worked per week, number of courses enrolled in along with the developmental mathematics course, transportation, family concerns, and other personal issues. Students must decide where completion of a developmental mathematics course fits in their many priorities. Students must also decide if they are willing to devote the time to the course if it fits in their schedules.

## *Chapter 3: Conclusion*

This pamphlet is intended to help developmental mathematics students answer many questions and make many decisions throughout their journey through developmental mathematics. This pamphlet by no means covers every possible question you might have, but the questions addressed here are very common based on my experience reading hundreds of responses about student issues during this journey.

It should also be noted that there are no recommendations here about academic ability or any specific developmental course or topic. There are many books that already address these issues. This pamphlet intends to address two significant nonacademic issues that have an impact on student success in developmental mathematics.

These questions should allow students to consider whether or not the time is right to enroll in developmental mathematics. Not all students are prepared for college, or, more specifically, for developmental mathematics. An attitude of willingness to learn needs to be in place prior to enrolling in developmental mathematics. As with other tasks, mathematics takes time. Students have to be sure they have the time to give before enrolling.

Following the advice in this pamphlet does not guarantee an A in a mathematics course, but it is very likely that a student who follows these recommendations will receive a passing grade. This is not to say that your goal in any course should not be to work hard for an "A" grade. In fact, an adjustment in attitude should tell you that you can do the work if you put your mind to it and remain focused. To add to this advice, understanding the attitude and time management strategies of other "A" students in developmental mathematics is also important. In conclusion, continue talking positive and believing that developmental mathematics is something that you can conquer at this point in your life. Make no more excuses, fight through the obstacles, and complete the journey!

Made in the USA
Monee, IL
13 September 2026

62977542R10020